A New Science
Worldview

Best Wishes

Don

A New Science
Worldview

A paradigm for better
living and working

DORN SWERDLIN

BALBOA
PRESS
A DIVISION OF HAY HOUSE

Copyright © 2012 Dorn Swerdlin

All rights reserved. No part of this book may be used or reproduced by any means, graphic, electronic, or mechanical, including photocopying, recording, taping or by any information storage retrieval system without the written permission of the publisher except in the case of brief quotations embodied in critical articles and reviews.

ISBN: 978-1-4525-6093-9 (sc)
ISBN: 978-1-4525-6092-2 (e)

Balboa Press books may be ordered through booksellers or by contacting:

Balboa Press
A Division of Hay House
1663 Liberty Drive
Bloomington, IN 47403
www.balboapress.com
1-(877) 407-4847

Because of the dynamic nature of the Internet, any web addresses or links contained in this book may have changed since publication and may no longer be valid. The views expressed in this work are solely those of the author and do not necessarily reflect the views of the publisher, and the publisher hereby disclaims any responsibility for them.

The author of this book does not dispense medical advice or prescribe the use of any technique as a form of treatment for physical, emotional, or medical problems without the advice of a physician, either directly or indirectly. The intent of the author is only to offer information of a general nature to help you in your quest for emotional and spiritual well-being. In the event you use any of the information in this book for yourself, which is your constitutional right, the author and the publisher assume no responsibility for your actions.

Any people depicted in stock imagery provided by Thinkstock are models, and such images are being used for illustrative purposes only. Certain stock imagery © Thinkstock.

Printed in the United States of America

Balboa Press rev. date: 10/17/2012

To my grandchildren,
Graham and Avery.

Contents

Preface	xi
Acknowledgements	xiii
Introduction	xv

Chapter 1 - Worldview — 1
    What is a worldview? — 1
    How to Expand Your Worldview — 2
    So how do we change our beliefs? — 3

Chapter 2 - Newtonian Thinking — 9

Chapter 3 - Quantum Thinking — 13
    Separateness — 13
    Determinism — 17
    Reductionism — 17
    Other Newtonian Concepts — 18

Chapter 4 - Consciousness — 21

Chapter 5 - Other New Science — 25
    Biology — 25
    Gaia Theory — 27
    Chaos Theory — 28
    Morphic Fields — 31
    Holographic Paradigm — 31

Chapter 6 - 2012 — 35
    Sunspots — 36
    Pole Reversal — 36
    26,000 Year Cycle — 36
    So What? — 37

Chapter 7 - Creating the New Science Worldview   41
   Consciousness   42
   Changing the Worldview   43
   Change Beliefs   43
   Change Thinking   43
   Change Worldview   44
   Consciousness   44
   Critical Mass   44
   New World   44

Bibliography   47

Preface

Although I cover many scientific concepts in this book, I am not a scientist. I am an actuary by profession and have had no formal training in physics, biology, or chemistry, except for basic courses in high school and college. What I know about these fields as well as "New Science" (to be defined later) I learned from reading books written for laymen.

My passion since my early 20's has been spirituality. Later in life a friend introduced me to quantum physics. We searched for books about science, and others about spirituality. We would then talk about what we read. It became clear to us that many scientific concepts have counterparts in spirituality, especially ideas in some Eastern teachings. They both describe the same phenomenon. This idea fascinated me.

In the early 1990's my wife Joanne and I attended an intensive leadership program presented by The Center for Authentic Leadership. In this program I learned more about science and spirituality as it applies to leadership. Around the same time my whole family, Joanne and our two children, Lee and Allison, learned to meditate. One of the leaders in the meditation program was Deepak Chopra, M.D. Dr. Chopra is a spiritual leader and has authored over 65 books. I was immediately attracted to his teachings because he spoke about new science and its connection with spirituality.

A few years ago I was certified as a Primordial Sound Meditation teacher at the Chopra Center for Wellbeing. My goal as a meditation teacher has been to teach employees of my firm to meditate. My reasons are twofold; to enhance the quality of life for our employees, and to improve the culture in our firm. As a business owner of an actuarial consulting firm, I have seen ways to improve our firm by thinking in a different way based on the science of the last 100 years.

In the 1996 election, I became involved with the national political party, The Natural Law Party (NLP). A main theme of the NLP is that our country can solve many of its problems if we study and follow the workings of nature and the universe as a whole. This theme keeps cropping up in my life and I go into more detail later in the book. I ran for Secretary of State in Georgia as a NLP candidate. The NLP was unable to have its candidates listed on the ballot. The issue went to court as we tried to have the NLP candidates listed on the ballot. The judge ruled against us. The existing two parties, in my opinion, work hard to keep third parties out. At least that was my experience. I didn't win. It wasn't close. I only picked up a few votes. Not that I would've won if my name was on the ballot. It was the beginning and end of my political career.

Acknowledgements

So many people have supported and helped me during my writing of The New Science Worldview. I would like to thank each and every one of you. I apologize in advance for those not mentioned.

Joanne, my best friend and wife, has put up with me for over 40 years. She has supported and encouraged me in so many ways, including editing the manuscript of this book.

My friend, Francis Klein, worked with me for years researching science and spirituality for the book. Francis also introduced me to Jan Smith of The Center for Authentic Leadership in Atlanta. Jan's programs taught me valuable life lessons and encouraged me to write my book.

Since the early 1990's I have followed Dr. Deepak Chopra. I have been attracted to him because of his ideas about science and spirituality. I've attended many programs at The Chopra Center and ultimately became certified as a meditation teacher. The co-founder of The Chopra Center along with Deepak was Dr. David Simon. David recently passed away and is sorely missed by so many lives he touched. I was fortunate to have had several medical consultations with Dr. Simon and watched how he helped so many people. Another leader in the Chopra Center is Davidji.

He encouraged me to pursue the teacher's path and become a certified meditation teacher. My own practice of meditation as well as my teaching experience has helped me to focus on this book.

Many years ago Dick McCormick worked with me as a consultant and coach. In addition to helping me with my business, he encouraged me to think about the genesis of my book. Randy Oven has been an advisor to me for about 20 years. We spent many hours discussing the progress of my book. Anson Ramsey, a friend for many years, is also writing a book. We met every month at my office to work on our books together. These meetings provided both of us an accountability for focusing on our books.

Thanks to Charles Stancil for his insightful review of my book from a scientific point of view.

The Leadership Management Team (LMT) at Swerdlin & Company has both supported me and encouraged me to take the time needed to focus on my book. The members of LMT are: Joanne Swerdlin, Lee Swerdlin (our son and president of our firm), Glenda Devechio, and Laura O'Connor. Kristin Hamilton at my office formatted the entire book.

Family members have been very supportive including, Allison Swerdlin, our daughter, Lee Swerdlin and David Golden.

I am grateful to all mentioned above as well as others not mentioned who have influenced my life and my book.

Introduction

This book is a short summary of our current paradigm, the Newtonian Worldview, and how it affects our world today. New Science provides a new paradigm and worldview which could minimize or eliminate many of our global issues. I summarize my version of this new paradigm. Let's begin by looking at some of the current issues within our organizations and throughout our world at large.

What's wrong with our organizations today?

- » Lack of creativity;
- » Focus on short-term profits to the exclusion of stakeholders such as employees, vendors, the community, and the world at large;
- » Gossip;
- » Bureaucratic rules and attitudes;
- » Withholding information at the top;
- » Greed;
- » Inflexible structure;.
- » Operates from scarcity beliefs;
- » Too much focus on competition, not enough on cooperation;
- » Operates from fear.

What's wrong with our world today?

Gregg Braden's book, *Deep Truth*[1], lists crisis points we are faced with right now:

- » Growing threat of war, nuclear war,
- » Growing shortages of food and water,
- » Climate change, and
- » Widening gap between:

  1. poverty and wealth,
  2. disease and health, and
  3. illiteracy and education.

I believe a fundamental cause of most of our issues today is the continuation of our worldview established about three centuries ago. I call this worldview "Newtonian Thinking." I explain what I mean by that term later in the book. For now let's say that our Western World has been under this "spell" for so long that it is well below our consciousness. We think this worldview is The Truth and the only Truth (with a capital T). We cannot change our worldview until we become aware of it.

It's time now to question this worldview and consider new paradigms because "Newtonian Thinking" not only limits our thinking, but also contributes to our current global problems. Our intuition and our senses clearly support and validate this outdated worldview.

In this book I attempt to show that Newtonian Thinking is obsolete in many ways, and how more recent scientific knowledge indicates this worldview is scientifically flawed in many aspects. I also propose specific ways to change our thinking from the Newtonian Worldview to the Quantum Worldview.

Chapter 1 discusses how our worldview affects our thinking and actions. I outline ways to change our worldview to align better with the knowledge we gain from the New Science.

Chapter 2 outlines the current worldview, I call Newtonian Thinking. Some of the concepts in this chapter include:

- » Separateness of things;
- » Determinism;
- » Reductionism;
- » No consideration of consciousness;
- » Time is linear, fixed, and independent of the observer;
- » Separateness of physics, biology and chemistry;
- » Paradoxes are ignored;
- » Evidence not fitting this worldview is ignored.

Chapter 3 discusses the new paradigm which I call "Quantum Thinking." This reflects new sciences discovered in the last 100 years or so. Quantum Thinking focuses on the physics discipline called quantum theory or quantum mechanics. Don't worry, I'm not a physicist and I will not get technical with the science discussion. I'll refer you to other sources if you are interested in delving deeper into the science and math.

Chapter 4 takes a leap from science into the world of consciousness.

Chapter 5 adds other "new science" discoveries which change our worldview. These include chaos theory, some new ideas in biology, morphic resonance, and the holographic paradigm.

Chapter 6 presents scientific and other views of the significance of the year 2012 and explains its potential effects on our planet and the human race.

Chapter 7 summarizes ways to expand our worldview beyond the current Newtonian Thinking.

Chapter 1

Worldview

What is a worldview?

In short, a worldview is a conceptual framework and a set of beliefs used to make sense out of a complex, seemingly chaotic reality. It becomes the source of our goals and desires, and as such, shapes our beliefs, behavior and values. Our worldview is so fundamental that we often are unaware of its existence as well as its total effect on our thinking. Being unaware of it prohibits us from considering it untrue and we become its prisoner.

Our worldview includes the common beliefs we share. A belief is an idea that we think is True, with a capital "T". We never consider its validity. Later I discuss ways I've found to test our beliefs. Barbara Marciniak in her book, *Path of Empowerment*[2] says:

> *"Beliefs are decisions and agreements you make about reality; they are the accumulation of invisible inner building blocks, formed from your*

interpretation of reality.... Beliefs are the programs from which you have built your life existence."

Your beliefs in total make up your worldview.

How to Expand Your Worldview

How do we change our current Western World thinking which is embedded in our culture? Our beliefs generate our thoughts and feelings which in turn dictate our behavior. Our reality is equal to the sum of our beliefs. Therefore, change your beliefs and change your world. Later we will talk about how to change your beliefs.

The quantum mechanical wave equation contains all possibilities, and we as observers, collapse the equation into one possibility (to be discussed in Chapter 3). However, we have no access to the possibilities we don't believe in. Your beliefs limit your worldview. Barbara Marciniak in her book, *Path of Empowerment*[3] says:

> "Until you put your beliefs aside, you will not see the world beyond your beliefs."

Our beliefs not only affect our thoughts, but they also affect our physiology. Sounds unbelievable? Dr. Bruce Lipton, a cell biologist, has conducted experiments which show changes in cell structure and DNA caused by changing beliefs. See his book, *The Biology of Beliefs*[4]. He challenges the traditional interpretation by biologists that our DNA is set at birth and remains constant throughout life. This changes the "nurture" vs. "nature" debate. This debate questions the relative effect on our lives, of:

1. Nature- our hereditary inheritance, vs.
2. Nurture- the environment and other factors affecting us during our lives.

Lipton maintains that our nature (DNA) actually changes, physically, as our beliefs (nurture) change. The nurture side of the equation becomes much more powerful as we add our own thoughts and beliefs to the list of external influences such as family and school. This idea of beliefs changing DNA is not yet recognized by the mainstream physics community.

This idea of a changing DNA shifts responsibility to us as creators of our own lives, rather than using our inherited characteristics as excuses for our problems. For example, "My father was impatient and I inherited it from him and that's why I'm impatient." This is a victim's excuse rather than taking responsibility for one's own choices.

Our beliefs create our perceptions and our perception is our reality. The old saying, "seeing is believing" is more accurately stated "believing is seeing", because what we see is always filtered by what we believe. Because of this, everyone lives in his or her own unique world. We often think others see things as we do. Not true.

In addition to beliefs which come from Newtonian Thinking, we also harbor beliefs derived from our ego which are often fear based. They are as harmful to our wellbeing as are beliefs derived from our Newtonian worldview. Our egos generate self-defeating inner dialogue, such as:

- » I can't deal with this situation,
- » I will always be poor,
- » Life is a struggle,
- » I've always done it that way,
- » This didn't work before, it never will,
- » This is a hostile and scary world,
- » I am helpless to change my life, etc.

As we repeat this dialogue to ourselves, it becomes embedded into our subconscious. We see the world as portrayed by our inner dialogue, reinforcing our beliefs.

So how do we change our beliefs?

One way that works is to become conscious of our internal thoughts. Start noticing what you say to yourself over and over again. One of the thoughts I repeat to myself is, "I am stupid". I usually say this when I forget something or make some kind of mistake. So begin to listen for what you say to yourself. In other words, think about what you think about.

Next you can ask the question "Is this really true? So, am I really stupid? Not really. I did achieve the highest credentials awarded to actuaries, Fellowship in the Society of Actuaries. Stupid people don't usually attain this level.

Then as you notice the negative dialogue, change it to something that is true. When I hear myself saying "I am stupid", I can change my thoughts to say "I'm not stupid just because I forgot something. Or "I may make stupid choices sometimes but that doesn't make me a stupid person." Or "using the word stupid about myself does harm to me."

Somewhere in my belief system I formed the opinion that I am stupid. I probably picked this up in my childhood. Repeating this phrase to myself reinforces my belief. I was not aware of this belief until I paid attention to my inner dialogue. I would not allow anyone to talk to my kids or grandchildren the way I talk to myself.

At my office, we organized a game to help us become more aware of what we think and say in order to improve our culture. We use this acronym: P-CRABLEG. Each letter represents a thought which is a cognitive distortion that can pollute our culture. A cognitive distortion is a saying or thought which is not true and/or is damaging to our communication with others.

Whenever an employee catches another speaking one of these distortions, the culprit must pay a quarter. In time we begin to catch our own distortions right after the words leave our mouths. In our game, no quarter is due if you catch yourself before someone else does. The ultimate goal is to catch yourself before you voice a distorted remark. We began thinking before saying something that could damage our environment. Communication has become smoother, cleaner and clearer as a result of our P-CRABLEG game. The acronym stands for the following:

P is taking things PERSONALLY. An example is we see someone with an unpleasant expression on his face. Our ego immediately kicks in and makes up a story like, "He must be mad at me. I wonder what I did to make him mad. He's got his nerve being mad at me. I didn't do anything to him…"and on and on. His facial expression most likely has nothing to do with you, but you have already created your own story as to why he is mad at you.

C stands for jumping to CONCLUSIONS. He came in late, he is a lazy slacker. This distortion also includes negative predictions, such as, "He'll never make it in this company."

R is RUMORS. "I heard the company will be laying off employees. Don't tell anyone I told you." These negative and fearful rumors spread like wildfire in an organization and pollute the culture.

A is for ALL or Nothing Thinking. These thoughts leave no gray areas and are only black or white. The following is an example of all or nothing thinking: "He came in late today, he is always late!" Whenever you see the words "always" or "never" it is likely you'll find all-or-nothing thinking. This example also shows jumping-to-conclusions as one day late means every day late.

B stands for BODY language. Body language can be more powerful than words and can block clear communication. In a meeting, rolling your eyes in response to another's comment will certainly clog up the communication lines. No words need be uttered. It is said that about 70% of communication is non-verbal.

L is for LABELING. Examples: "She is a drama queen." "This office is a sweat shop".

E stands for EXAGGERATING. "I've told her a thousand times." Did you tell her twice or 8 times or what? This kind of exaggeration leaves the listener with inaccurate, misleading, or insufficient information.

G is for GOSSIP. How much time and energy is wasted on idle gossip? Efficiency, office atmosphere, and people's attitudes are all polluted by gossip.

Another question in this process of changing your beliefs is to see how your behavior might be affected by your belief. Are there patterns in your behavior that might come from this belief? One of the negative inner dialogue examples mentioned earlier is "This is a hostile and scary world." A person with this belief lives in a hostile world. He or she might be overly suspicious of others, and may assume that everyone is out to hurt them. With this belief one's world *is* scary. People tend to attract others with similar negative beliefs. The point is your experience of a scary world is an indication that you believe the world is frightening. With the opposite belief one can enjoy a safe and nurturing world.

The above ideas about changing your beliefs were developed in part from my reading of Wayne Dyer's book *Excuses Begone*[5].

I like the views expressed about beliefs by Barbara Marciniak[6] in her book mentioned previously:

"Everyone has a bushel basketful of beliefs collected from many sources. Generic inheritance, family proclivities, childhood experiences, social and cultural influences…. Beliefs are the thoughts held, most often without question, about yourself and the world at large. These unnoticed thought-forms are birthed in your subconscious memory bank, qualifying your experiences in both the inner and outer worlds. Your beliefs set you up for success or failure. If your worldview is optimistic, you will generally be self-motivated with a positive attitude and engage life with joy and enthusiasm. —People suffer from pain and confusion due to a belief in their own sense of powerlessness-— Your encounters in the outer world are a reflection of your inner reality: you become what you think about; therefore, when you change your thinking, you will inevitably change your life. "

Another quote provides the first step in changing your beliefs:

"One condition that must be met before we can unleash the power of belief: We must believe in belief itself for it to have power in our lives."

This comes from the book, *Spontaneous Healing of Belief* by Gregg Braden[7]. So we have to believe that beliefs work to change and guide our lives. If we don't, it's much less likely to work.

I find interesting the following quotes from the book, *The Fifth Agreement*, by Don Miguel Ruiz and Don Jose Ruiz[8].

> *"Our belief system is just like a mirror that only shows us what we believe."*

The authors of this book talk about how our world is created by us from our beliefs. They distinguish the virtual reality from the real reality. The virtual reality is the reality we create in our minds. It is our interpretation of the actual "real" reality and thus becomes our world. Each of us has a different world because our beliefs are different. According to quantum physics, reality is a realm of possibilities and our specific observation of this reality creates our world. An analogy used in *The Fifth Agreement* is that when we look into a mirror, we see many objects. When we try to touch the objects in the mirror, we find that we only touch the surface of the mirror. The mirror is a virtual reality reflecting the "real" reality. So our perception of the world outside of us is a virtual reality residing inside.

> *"We make the assumption that what we believe is the absolute truth, and we never stop to consider that our truth is a relative truth, a virtual truth."*

A quote from *The Path of Empowerment*[9] adds another way to change your beliefs:

> *"…feelings take you back to beliefs, so find the feeling and really feel it. Once you have identified the belief behind the feeling, acknowledge the role of the belief and replace it with a more empowered outlook."*

Speaking of beliefs, the following are some of mine:

- » I believe in a higher power; from a religious point of view, I'll call this power God. In scientific terms, I'll say the Unified Field. Another moniker is Pure Consciousness,
- » I believe in reincarnation, (as an actuary, I studied mortality, but now I am interested in immortality),
- » I believe this higher power is the source of our universe as well as the multiverse (infinite universes),
- » I believe consciousness is primary to matter and energy. That is, matter and energy manifest from consciousness.

- » I believe that humans have much greater abilities than our Western Worldview tells us,
- » I believe we have the technology to fix all of the world's problems if we accept our responsibility and ability to make these changes,
- » I believe the way to change the world is to change ourselves,
- » I believe a small minority of the population (the critical mass or tipping point) can change the whole world,
- » I believe that observers manifest energy and matter from the nonlocal quantum realm.

Chapter 2

Newtonian Thinking

Isaac Newton in the late 17th century pulled together ideas first presented in the world by Nicolaus Copernicus, a French scientist and philosopher. Copernicus introduced the idea that the earth revolves around the sun, rather than the earth being the center of the solar system and the whole universe. The paradigm at the time assumed that the earth stood still and all celestial bodies moved around us. The proposed new idea certainly is counterintuitive. Critics at the time said, "If the earth is moving, we would certainly blow off the planet, or "It is obvious that the sun revolves around the earth because we can see the sun moving every day." This was a revolutionary concept which the world was not ready to accept.

Newton's demonstration through mathematics (using calculus he invented) shows that the laws of motion of the planets, the moon and stars are the same as those on earth. This was the first major scientific discovery which generated unity. Newton unified earth and the heavens. Subsequent scientific breakthroughs further unified laws of nature. For example, in the late 1800's Maxwell integrated the two separate forces at the time; the electrical forces and magnetic forces into a single force called electromagnetic energy.

Einstein unified energy and matter with his famous equation $E=mc^2$ (Energy equals mass times the speed of light squared). He also united space and time into one and called it space-time. The next unification combined electromagnetism with the weak force to be called electroweak. The weak force is associated with radioactivity.

Currently physicists are working to develop a unified field or "theory of everything" which would unify and integrate all known forces into one mathematical theory. This accomplishment would unify gravity, electroweak and the strong forces. The strong force holds the nucleus of atoms together.

Back to Sir Isaac: Our current worldview, established by Newton looks like this:

1. Things are separate. Things only move by outside forces such as gravity or magnetic forces. The only connection among things is outside forces.
2. Determinism. If one could "see" all the forces working in the universe then one could predict the future in a clockwork fashion, like a big mechanical clock. The universe includes predictable outcomes based on Newton's clockwork view.
3. Reductionism. This concept assumes that if we find the component parts of a complex system, then we can learn about the whole. Matter can be broken down into smaller and smaller parts and scientists could ultimately identify the smallest pieces of matter. The atom was thought to be the smallest particle of matter in Newton's time. We now know about many smaller subatomic particles such as protons, neutrons, quarks, and others. I first learned about subatomic particles watching the 1950's tv show, Amos & Andy, when Kingfish was teaching Andy saying: "the proton, neutron, fig newton and moron".
4. No consideration of life processes or consciousness is included in classical physics. Science and religion (and spirituality) were separate and distinct. Adding to this separation was the

religious politics of the time where the Church did not like its beliefs challenged, and the resulting threat to its power. Most current physicists still consider consciousness outside of physics, however, an increasing number are starting to think about the deeper meaning of physics and how consciousness may play a part.
5. Time is linear, unchanging, and an independent phenomenon.
6. Physics, chemistry and biology are separate disciplines with little coordination,
7. Paradoxes are ignored, and
8. Evidence not fitting into the current paradigm is ignored. This is typical in any paradigm shift.

It is important to note these concepts derived from Newton's laws became the way of thinking, but Newton himself was a spiritual man believing in a higher power orchestrating the movements of the universe. He also worked to see the relationship between science and religion. He learned the Hebrew language to look for a code in the bible. His creation of the science and the math to work with his science is still used today. At the macro level his science still works well. Our conquest of space, for example, uses only Newton's science. However his science breaks down at the atomic and subatomic scales.

In spite of the value of his works, our behaviors based on the concepts inherent in his science no longer serve us.

Notice how all of these concepts seem very logical and unquestionable. We accept these as obvious and fundamental principles that we live by and would never question. Our intuition and senses confirm these ideas. This belief shapes our worldview.

Why should it be otherwise? I'll tell you why; modern science has proven scientifically that each of the above precepts of "classical physics" is no longer true. It's time to look at these proven ideas and begin living in accordance with them. Stanislav Grof eloquently summarized this concept in the following quote:

"Western science is approaching a paradigm of unprecedented proportions, one that will change our concepts of reality and of human nature, bridge the gap between ancient wisdom and modern science and reconcile the difference between Eastern spirituality and Western pragmatism".

We dig deeper into this current paradigm shift later in the book.

The next chapter outlines the "corrections" in Newtonian Thinking, proven by more recent science.

Chapter 3

Quantum Thinking

I use the term "Quantum Thinking" to represent the new way of thinking, underlying a new paradigm and worldview based primarily on the science of quantum theory. How does quantum theory change the Newtonian conclusions discussed earlier?

I must warn you that many of these concepts may sound strange and seem weird and unbelievable. Even the scientists who were instrumental in developing quantum theory say the results are weird. Neils Bohr said, *"Anyone who is not shocked by quantum theory has not understood it."* Also, Erwin Schrodinger announced, *"I don't like it and I'm sorry I ever had anything to do with it."*

Let's look at some Newtonian concepts as compared with Quantum Thinking.

Separateness

Newtonian Thinking looks at things as being separate. Quantum Thinking focuses on relationships rather than separate "things". I believe this concept of separateness is a major flaw in the current Western Worldview. Many of our problems could be eliminated if we change this belief.

Our perception of matter in the world leads us to believe that things are separate. My body is separate from your body and the rest of the environment. Is this "true" or only our perception of the world? Of course our perception is our reality and therefore "true" for us. Science has proven that at the subatomic level, where the quantum realm "resides", everything is interconnected with everything else. Another term for this nonlocal interconnectedness is "quantum entanglement".

One characteristic of this realm is "non-locality". That means that nothing has a place in time or space. Everything is at once everywhere and nowhere. I told you these concepts are weird. How could anything be everywhere and nowhere? How can anything not have a place in time? Although it's difficult to even conceive of non-locality, science has proven its existence beyond a doubt. This is just another example of science uncovering phenomenon which clashes with what we experience and believe. Earlier examples include the discovery that the earth is round rather than flat and that the earth revolves around the sun instead of the other way around. This non-local realm is at the subatomic scale. Everything at this level is connected and there is no separateness.

The illusion of separateness is also evident when we look closer at our macro material world. We see our bodies as separate from our environment. In reality we continually exchange material between our bodies and our environment. Each breath we take in brings oxygen into our system. Each time we exhale we expel carbon dioxide into the world. As our cells die they are replaced with new cells from molecules outside of our body. For example, the molecules in our liver are completely replaced every six weeks. Our skin molecules are replaced every month. Even our brain cells are completely different after one year. Our physical bodies are in constant dance with the environment. We don't see the change, but the material that makes up our physical body changes continually. So the boundary between our body and our environment is illusory. Our perception and our beliefs about our separateness are not consistent with the reality of the actual exchange of material. The Newtonian Thinking concept of separateness does not tell the real story.

Our bodies appear to us to be solid and static. Obvious to us is a child's body changing rapidly through the growth years. Less obvious is our slower changes as adults. Our 65 year old body is clearly different from our 25 year old body. Our body is more like a river with energy, matter and information flowing in, out and through us, rather than a static structure like a machine. Western medical science treats us like machines with parts to fix when broken, rather than integrated systems in constant flow with the environment. This is why our doctors are so specialized. Medical specialists generally focus on parts of the body rather than the body as an entire system. These specialists know a whole lot about their area (heart, lungs, feet, skin, etc.) but not so much about the relationships and connections among the other parts.

Our pharmaceutical providers develop a drug to deal with a specific condition. They often miss the unintended consequences because they don't see the connections between the parts of our bodies. The side effects of drugs represent some of these unintended consequences. For example, drugs being recalled because of harmful or fatal reactions. Another example is when a drug is developed for one condition and later is used for a different purpose. In both cases the focus is on one separate symptom, and later we find other parts of our body affected. Our bodies do not operate as separate parts in one machine, but as an integrated system where everything affects everything else sort of like the non-local realm from quantum mechanics. The non-local realm serves as a much better model of our bodies than the mechanical model used in Western science.

Our separateness thinking generates an "us vs. them" mentality. We are separate and we are different. You are different and I know my way is good, so yours must be bad. My god is the right god and your god is wrong and I must destroy you to protect myself and defend my god. How many people have been killed in the name of various gods in our human history? And after all these millennia it seems to be getting worse rather than better. Only as recent as the middle of the 1900's we committed massive genocide. Racial and religious killing and terrorism continues today in a big way with no sign of improvement in sight.

If we had an understanding of our connections and similarities with others rather than our separateness and differences, we might be less likely to hate and fear others. If we believe that hurting others also hurts us, we might act differently. If we know we are connected we wouldn't "cut off our nose to spite our face." The tactic of "divide and conquer" and the concept of "united we stand, divided we fall" both indicate the downside of separateness.

Another result of our separateness is our relationship with our mother Earth. Most have experienced the calmness and serenity we feel in a beautiful natural setting. We can often feel the connection with nature. Newtonian Thinking leads us to believe that we are separate from our beautiful home planet. Quantum Thinking tells us we are connected to our planet, nature and other living things. I consider our earth a living organism rather than a big lifeless rock with some life on its surface. I'll talk further about this hypothesis (Gaia theory) later.

Albert Einstein provides a comment on separateness in the following quote:

> "A human being is part of a whole, called by us the Universe, a part limited in time and space. He experiences himself, his thoughts and feelings, as something separated from the rest....a kind of optical delusion of his consciousness. This delusion is a kind of prison for us, restricting us to our personal desires and to affection for a few persons nearest to us. Our task must be to free ourselves from this prison by widening our circles of compassion to embrace all living creatures and the whole of nature in its beauty."

I especially like Einstein's phrase, "optical delusion".

One of the most valuable concepts, in my opinion, is the quantum idea of "observer effect". It also clashes with the Newtonian idea of separateness. Quantum physics proves our conscious focus affects not only what we see, but also affects what actually manifests into reality. In a famous experiment called the "double-split experiment" electrons are fired through a single slit and then a double slit and its paths are recorded on a screen. The electron through the single slit hits the screen as would a particle. When you add the second slit, one would expect the electrons to hit the screen at two separate points on the screen from the two separate slits. To the surprise of the experimenters, the screen from the double slit shows clear indications that the electron is a wave! This is very counterintuitive to me. Isn't an electron either a wave or a particle? How could it be both at the same time?

So what does this experiment have to do with the observer effect? When the unexpected wavelike aspects of the electron showed up, the scientists decided to look at only one of the double slits to see what would happen. They used a measuring device to "observe" the slit. Surprise! When measured or observed the wave becomes a particle. The act of observation collapses the spread-out wave into a pinpoint particle. Mathematically the wave is expressed in the Schrodinger wave function representing a probability distribution. The wave function gives the probability of finding a particle at a certain position.

Max Planck, one of the founding fathers of quantum theory, provides a quote about the observer effect:

"All matter originates and exists only by virtue of a force of an atom to vibration which holds the atom together. We must assume behind this force is the existence of a conscious and intelligent mind. This mind is the matrix of all matter."

Determinism

Under Newton's laws the universe works like a really big machine with its separate parts working in an orderly fashion. Predictability is possible, with sufficient information. Quantum theory shows us that at the quantum realm anything is possible and nothing is predictable as a certainty. Furthermore, at this level, only probabilities exist and in mathematical terms this realm is represented by a probability wave function; the Schrodinger Wave Equation.

Reductionism

Reductionism is a Newtonian concept which assumes if we can find the component parts of a whole, then we can better understand the whole. This concept has lead physics on the road to finding the ultimate smallest pieces of matter to help explain the workings of the universe. However, quantum physics shows that at the most fundamental level, only probabilities exist and locating subatomic particles as a fixed solid is impossible. An example of reductionism comes from Rene Descartes, a scientist in Newton's time:

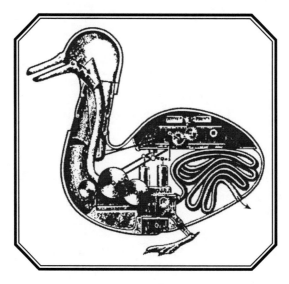

Descartes held that non-human animals could be reductively explained as automata (robot like). In other words, he assumed you could build a duck with only the requisite body parts

OTHER NEWTONIAN CONCEPTS

Newtonian science assumes that time is fixed and independent of outside influence. Einstein proved that time is a perception and varies with observers traveling at different speeds. He also showed that space and time are aspects of a combined entity called space-time. This new concept of time is very counterintuitive because our senses do not perceive time as it really is. We experience time as linear and constant because that's what our beliefs tell us. The time distinction does not become significant until the observer reaches speeds near the speed of light (671 million miles per hour). That's fast! Time is a varying perception rather than a fixed cosmic constant. This is one example of how our perception is our reality, and each of us has a unique reality. That is, reality for each of us is created from our personal thoughts, beliefs and worldviews. Since we have the ability to change our thoughts and beliefs, we are empowered to change our world. We have a lot more power than our Western worldview allows.

The Newtonian paradigm uses "either-or" thinking while Quantum Thinking adopts the "both-and" approach. As an example, the old paradigm says that nothing could be both a particle and a wave. It is either one or the other. Quantum Thinking shows that an electron, for example, can be both a wave and a particle. This example shows that the particle-wave duality is a paradox in Newtonian Thinking. Paradoxes are ignored. However, Quantum Thinking has many paradoxes.

Another characteristic of Newtonian Thinking, or any paradigm shift, is the tendency for those imbedded in the old paradigm to ignore evidence not fitting the current thinking. The next chapter expands on this idea, as well as discussing consciousness in science and beyond.

Chapter 4

Consciousness

To me, one of the most significant changes from Newtonian Thinking to Quantum Thinking is the introduction of consciousness in Quantum physics.

As mentioned above, many established physicists do not accept the idea of consciousness deriving from quantum mechanics. I've talked to physicists who consider consciousness to be beyond science and any reference to spirit or higher powers as not science but as philosophy or religion. This is typical of new paradigms. The insiders (physicists) tend to be the last to accept ideas not includible in the current paradigm. Outsiders usually identify inconsistencies in the current paradigm and can be more open to change. This is not a criticism of physicists because it's human nature to believe what you have learned and taught over the years. It is also difficult for a practicing physicist to respect a non- physicist's opinion in their area of expertise.

Gregg Braden in his book *Deep Truth*[1] addresses this issue of new discoveries:

> "While for some people the possibilities hinted at by new discoveries are a refreshing way to view the world, for others they shake the foundation of long-standing tradition....It's sometimes easier to rest on the false assumptions of outdated science than to embrace information that changes everything we understand. When we take the easier course, however, we live in the illusion of a lie."

Now I would like you to bear with me as I ascend into a realm beyond science. As a Primordial Sound Mantra meditation teacher, I teach the 7 levels of consciousness, as described by Deepak Chopra, based on ancient Vedic teachings. Many other spiritual disciplines outline similar states of consciousness. The Vedic view lists the seven levels of consciousness as:

1. Deep sleep,
2. Dreaming sleep,
3. Waking state,
4. Glimpse of the soul or the transcendental state,
5. Cosmic consciousness,
6. Divine consciousness,
7. Unified consciousness.

In our deep sleep state we have little consciousness, but we are conscious. We are aware enough to hear a loud sound, for example.

In our dreaming state we see ourselves, other people and other things. In this state all of the activities are in our head, but we experience separation between us and others. Here's the interesting part: In this state the dream is our reality. As a matter of fact in every state of consciousness we experience a different reality, and our physical bodies are different in each stage of consciousness. Our brain wave frequencies change as we expand our awareness. This has been scientifically documented.

The third state, the waking state is where we live every day. When we remember our dreams we say that our dreams are not real, it was just a dream. However it was real to us while we were dreaming. The dream is our reality while we dream. As we ascend into higher states we are aware of the "lower" states. The first example is our remembering our dreams (second state) while we are conscious in the waking state (third state). The shift here is to realize that dreaming is real rather than being just a dream that is some kind of an unreal mental process. What we consider as the only reality is shown to be our reality ONLY in the waking state. That is, there are realities other than what we experience in our waking state. Life is much richer than it appears to be. We have capabilities way beyond our Western World viewpoint.

At the fourth state of consciousness we begin to glimpse into the nonlocal realm. This state is often called the transcendental state of consciousness because our awareness goes beyond our waking state and beyond space-time. We all experience this awareness at certain times, as when we observe a beautiful sunset. We may have a feeling of connection with nature or even the entire universe.

The 5th, 6th, and 7th states of consciousness bring our awareness to a more and more expansive perception level. At each level our reality and physical bodies change. At each level we perceive all lower levels.

While we are in the mode of reaching beyond science, I refer to a book *Bridging Science and Spirit*[2], by Norman Friedman. He postulates that units of consciousness or CUs exist as the fundamental source of our three dimensional world.

In *My View of the World*[3], Schrodinger writes:

"Consciousness is that by which this world first becomes manifest, by which indeed, we can quite calmly say, it first becomes present; that the world consists of the elements of consciousness..."

I teach the 7 levels of consciousness in my meditation class because I believe that meditation is one way to expand consciousness. An expansion of consciousness by enough people is our way to deal with the current ills in our world. David Bohm, a famous physicist, says:

"As you probe more deeply into matter, it appears to have more and more subtle properties...In my view, the implications of physics seem to be that nature is so subtle that it could be almost alive or intelligent."

I believe that the whole universe is alive and conscious. As we expand our own consciousness we begin to become aware of the consciousness of everything including inanimate objects.

Chapter 5

Other New Science

Biology

In the 20th century the discipline of physics enjoyed a significant expansion of basic views. The Newtonian thinking or materialism was replaced by new ideas from Quantum theory. Biologists, however, remained stuck in the materialism of Newton's science. Rather than introducing concepts of consciousness and other outside forces on matter, biology focused only on the molecules of life. Bruce Lipton, a cell biologist introduced in chapter 1, once read a layman's book on quantum physics. He was fascinated by the concepts and looked deeper into the new physics. He realized that in order to fully appreciate the science of biology, one must first understand the more fundamental science of physics.

A major factor in the lack of biological expansion during the 20th century was the influence of drug companies. These companies enjoyed significant profits by creating pills. The underlying assumption is that pills (material molecules) can alone cure illnesses. Quantum ideas, which change this assumption to include outside and invisible forces, were down played by the business forces to produce pills and focus only on molecules. At the same time, paradigm busters such as Bruce Lipton, Rupert Sheldrake, Deepak Chopra, and many others, incorporated the new physics into the study of biology.

Deepak Chopra outlines an interesting analogy in his book, *The Book of Secrets*.[1] He lists the characteristics of the cells in our body and shows how we could improve our lives if we act as our cells do. This is another great example of how life provides a roadmap to follow to improve our lives. Here are some characteristics of our cells according to Dr. Chopra:

- Higher Purpose. Every cell agrees to work for the welfare of the whole body; its individual welfare is secondary. Cells even die for the welfare of the whole body. For example, each month every skin cell dies and is replaced by new ones. ***Selfishness is not an option.***
- Creativity. Although every cell has a unique function, cells can combine in creative ways. A person can digest food never eaten before or think thoughts never thought before. ***Clinging to old habits is not an option.***
- Efficiency. Cells operate using the smallest expenditure of energy. ***Excessive consumption of food, air or water is not an option.***

So how can we apply these concepts to our lives?

Higher purpose. If we all lived with the same higher purpose as our cells, we would have no hate, no wars, no hunger, no terrorism, etc. Everyone in an organization would put the company ahead of their own personal gains. No unions would be needed; no regulations would be needed to control our behavior at work. And the list goes on.

Creativity. We humans are wired to live by habits and they are very useful at times. For example, when we drive our vehicles, our habits allow us to drive without thinking of every move. We automatically drive the same path to work without thinking about every turn. However, if we allow our habits to run our lives, we lose the conscious control of our choices and actions. We become prisoners of our past and we abandon our creative abilities.

Efficiency. Cells never consume more than they need. If we could do this, we would have no obesity, no energy issues, no greed and no hunger. With this maximum efficiency, our businesses would be more productive and more profitable.

Think of the world operating based on these three characteristics of our cells. Our major global problems would disappear. Nature works quite well, with no effort. If we adopt some of nature's ways of being and evolving, our world would certainly be a better place to live.

Gaia Theory

Another recent biological theory which I include in the "New Science" category is known as Gaia theory or the Gaia hypothesis. The name Gaia comes from the Greek goddess of the Earth. In short, the theory assumes that our earth is a living organism rather than a big rock with life on its surface. The big rock idea is the current paradigm.

Sir James Lovelock, a British atmospheric scientist and chemist, consulted with the space agency, NASA in the 1960's to address this question, "How do we determine whether or not life exists on Mars?" This leads to the more basic question, "How do we define the line between life and non-life?" In his research for NASA, Lovelock created the Gaia hypothesis; that our earth itself is a living being.

Until recently we had no reason to imagine our planet as an organism. Peter Russell in his book, *The Global Brain*,[2] compares fleas living on an elephant with us earthlings living on earth.

> "They chart its terrain...study it's chemistry, plot its temperature changes- arriving at a reasonable perception of where they live. Then one day a few of the fleas take a huge leap and see the elephant from a distance and see that it is alive."

Another reason we haven't thought of the earth as a living being is that we don't experience its changes over time. Our time frames are very different. Russell says, "If we could speed up time appropriately, we could see the atmosphere and ocean currents swirling around the planet, circulating nutrients and carrying away waste products, much as our blood circulates nutrients and carries away waste in our own bodies."

When our astronauts saw the earth as a "blue pearl" hanging in space (like the fleas hopping off of the elephant), their connection with mother earth caused a profound shift in their beliefs about our world. The sixth man to stand on the moon, Edgar Mitchell, is quoted as saying, *"it was beautiful, harmonious, peaceful looking planet, blue with white clouds, and one that gave you a deep sense... of home, of being, of identity."* The first picture of earth from space was shown on the cover of Life magazine, January 10, 1969. Another point of view comes from anthropologist Margaret Mead who called this image–

> *"the most sobering photograph ever made. Our lovely, lonely planet afloat in a vast black sea of space. So beautiful yet so tragically fragile. So dependent on so many people in all countries."*

CHAOS THEORY

Although not directly part of biology, another member of "new science" is Chaos theory. The Greek goddess Gaia had a partner, the Greek god Chaos. Together they create all that is. Chaos is the dark void from which mother Gaia brings form and substance. The two represent the duality in our world, including the creative force of male and female in our biology.

So what is Chaos theory? First let's look at how the theory was developed. In the 1960's, Edward Lorenz, a meteorologist, was studying weather patterns with a complex computer program. At least it was complex in the early 60's. To save space, he rounded some input data to 3 decimal places (.506) even though the computer stored data to 6 places (.506127). The results with this minor change turned out to be wildly different than the results from the original 6 decimal calculation. Lorenz knew this was impossible, and he assumed he had entered the data incorrectly. He double checked and found he had entered the data correctly. This accidental discovery of this very small change in the initial input, generating amazingly different results, became the genesis of chaos theory.

Lorenz coined the phrase, "butterfly effect" to represent this discovery. He gave a talk in 1972 entitled, *"Predictability: Does the flap of a butterfly's wings in Brazil set off a tornado in Texas?"*

According to the Newtonian paradigm small differences are assumed to be immaterial and are ignored. Lorenz discovered that with nonlinear systems, such as weather patterns, small changes in initial amounts can generate very different results.

Benoit Mandlelbrot, a French mathematician, was a major contributor to Chaos theory. With the help of modern computers, Mandlelbrot used a simple iterative formula; i.e. the answer to the equation is plugged into the same formula and repeated over and over again. He ran the iterative process millions of times and plotted the results. The picture arising from the graph at first was random and chaotic, but after much iteration a pattern emerged. The same patterns appeared at various "scales" no matter how many times the formula was repeated. They call this "self-similar". The big surprise was that the emerging patterns starting from various simple formulas looked like trees and bushes, our lungs, snails, clouds, mountains, and on and on. These patterns reveal a whole new insight into how life works; start with a simple formula, iterate over and over with feed-back loops, and tree limbs; our lungs, our finger nails, the clouds, and many other natural entities emerge. Patterns also emerge which duplicate stellar and galactic shapes. It appears that order to chaos to higher order is the process of life as well as the entire universe.

In their book, *Seven Life Lessons of Chaos*,[3] John Briggs and F. David Peat, both physicists, provide insightful uses of Chaos theory in our daily lives. I summarize below, in my own words, some of their ideas:

Creativity: The book describes creativity as not a special talent reserved for a few, but rather a mindset; a mindset which forfeits the constricted grip of our ego, our fear of mistakes, our pull to stay in our comfort zones, and our fear of the unknown. Chaos is where the unknown resides, and where the creation of new order begins.

Nonlinearity: In the Newtonian linear world, small initial changes result in small changes in a system. Chaos theory shows that nonlinear systems can generate very significant changes in the system from a very small initial change. This is the "butterfly effect" discussed earlier. The difference between linear and nonlinear can be as much as the difference between 10*10 (100) and 10^10 (10,000,000,000). The former is linear and the later is exponential. An example in the book is Rosa Parks, who by the small event of one person refusing to sit in the back of the bus resulted in thousands of people boycotting buses, leading to the eventual fall of segregation.

In an organization, an innocent comment can blossom into a major breakdown of the entire company's culture.

Newtonian thinking considers disorder as a destructive force. Entropy is the term describing the process of disorder and decay. Newtonian Thinking depicts the world as decaying into disorder. Chaos theory shows us that disorder is the way to achieve a new and higher order. In organizations we work very hard to keep order. Order keeps us stuck in the past with little hope of attaining a higher order.

Chaos theory brings many practical applications. In my business, I have applied Chaos concepts to help my firm prosper. Looking at how feedback loops work within Chaos theory, I have seen how one comment, especially from our management team, can blossom into a major issue with all the melodrama that follows. Newtonian linear thinking would lead us to believe that an innocent comment would not cause a big disturbance in the firm. Chaos theory tells us otherwise.

On the other hand, there are times when "stirring the pot" is appropriate. Shake up the status quo. An example comes to mind where an employee explains why a certain process is used. Quite often the answer is "we've always done it that way". This is a good time to stir the pot by responding, "Let's see if we can find a better way and forget about what was done before". This disturbs the status quo and opens the door to creative thinking. Chaos theory says that new order comes from the old order in a state of chaos.

Morphic Fields

Rupert Sheldrake, a famous biologist, developed a theory called morphogenetic or morphic fields. Sheldrake says in one of his articles on his website: "*These morphogenetic fields work by imposing patterns on otherwise random or indeterminate patterns of activity.*" An example often used in describing this type of field is called "The 100th monkey effect." This story tells of monkeys on a beach eating sweet potatoes. The monkeys had to remove the sand from the potatoes before eating. It was not easy to clean them. Then one day a monkey washed his sweet potato in the surf to easily clean off the sand. The other monkeys on the beach learned the new cleaning method and when the 100th monkey washed his potato, instantaneously all members of the species all over the world began washing potatoes. The 100th monkey represents the critical mass or the tipping point.

Continuing with Sheldrake's quote; "*Morphogenic fields are not fixed forever, but evolve. The fields of Afgan hounds and poodles have become different from those of their common ancestors, wolves. How are these fields inherited? I propose that they are transmitted from past members of the species through a kind of non-local resonance called morphic resonance.*" He also applied the field to humans: "*The fields organizing the activity of the nervous system are likewise inherited through morphic resonance, conveying a collective, instinctive memory. Each individual both draws upon and contributes to the collective memory of the species.*"

Newtonian science does not recognize this field and assumes all fields in nature are constant and eternal.

Holographic Paradigm

In 1982, Alain Aspect, a physicist at the University of Paris, discovered that subatomic particles such as electrons, under certain circumstances are able to instantaneously communicate with each other regardless of the distance between the particles. They could be 10 feet or 10 billion miles apart. This interaction is called quantum entanglement. Einstein didn't like this concept and called it "spooky action at a distance". This discovery may be one of the most significant discoveries of the 20th century. The reason:

If particles can communicate instantly regardless of their distance apart, Einstein's requirement that nothing can travel beyond the speed of light appears to be violated. Holograms can address this contradiction.

My first introduction to holograms was in a movie theater while watching *Star Wars*. The robot named Artoo Detoo shoots out a beam of light and magically projects a three dimensional image in the air. It is a small image of Princess Leia, asking for someone named Obi-wan Kenobi to come to her rescue. The image is a hologram. A hologram is created by shining a laser beam on an object. Then a second laser beam is bounced off the reflected light of the first and the resulting interference pattern (the area where the two laser beams collide) is captured on film. The developed film looks like a meaningless chaotic pattern. However, shining another laser beam on the developed film creates a three-dimensional image of the original object appearing to float in the air.

Another characteristic of a hologram, more amazing than the three-dimensional image, is what happens when the hologram is cut in half. Rather than each half containing one half of the image, each half contains the entire image, smaller and fuzzier than the original image. As you cut the image over and over again each piece contains the entire image.

David Bohm, a physicist, was one of the primary proponents of the holographic paradigm. He provides an illustration to help visualize this phenomenon:

> Imagine an aquarium with one fish swimming in it. Two cameras point to the aquarium from two different angles: one directly to the front and the other camera points to the side of the aquarium. Imagine that we cannot see the cameras and we only see the two images from the cameras. We see two separate fish swimming in two different fish bowls. We notice that they move in different directions, but we notice that as one moves so does the other. They seem to be communicating with each other. In reality there is only one fish. Our perception of two particles separated by great distances acting in unison is an illusion same as the fish story.

This explains the apparent faster than light communication, and no conflict with Einstein's speed of light limit.

This is another discovery which violates the Newtonian concept of separateness. Bohm concluded that there must be an underlying, deeper level of reality from which matter is manifest.

Stanford neurophysiologist Karl Pribram was interested in finding where memory is stored in the brain. He realized that the brain works like a hologram as the memories are dispersed throughout the brain. Pribram believes memories are encoded not in neurons but in patterns of nerve impulses that crisscross the whole brain. This is the same way that patterns of laser light interference crisscross the entire area of a piece of film containing a holographic image. So the brain itself is a hologram. And then under the holographic paradigm the whole universe is a super big and incredibly complex hologram. Talk about counterintuitive! If you consider this to be true, your worldview will surely shift away from Newtonian Thinking.

The Western world has traditionally assumed that the brain generates consciousness. The holographic paradigm purports that consciousness creates the appearance of the brain. That is, consciousness is the underlying reality while matter, including the brain, is manifest from this consciousness.

The next chapter highlights the significance of the current time frame.

Chapter 6

2012

When something is transformed it changes into a different form and can never go back to its original state. A good example of transformation is the caterpillar transforming into a butterfly. The butterfly is certainly different from its caterpillar source and it cannot go back to being a caterpillar. My family is experiencing transformation now. My granddaughter, about 9 months old at this writing, was transformed from a single cell to a beautiful little baby girl.

I believe the planet and the human race is ready for a transformation of unbelievable significance. I discuss some scientific evidence of this transformation below.

This year's winter solstice is the end of the calendar calculated by the Mayans about 1,500 years ago. Many different theories and prophesies have commented on the significance of December 21, 2012.

Sunspots

Scientists have analyzed sunspot cycles since 1610. Sunspots are massive magnetic storms on the sun. The analysis showed that since 1610 there have been 23 cycles averaging about 11 years each. The last cycle was 1996 to 2006. On March 10, 2006 the sunspots stopped abruptly, ending the cycle. History indicates that the end of a cycle is an indication that a new cycle is beginning. The cycles are expected to hit a maximum level this year. Muasumi Dikpati of the National Center for Atmospheric Research (NCAR) says *"The next sunspot cycle will be 30-50 percent stronger than the previous one."*

Pole Reversal

A pole reversal is when earth's magnetic poles actually shift 180 degrees so that the north and south poles' magnetic fields are exchanged. Geological records show that pole reversals happen routinely throughout the earth's history. Magnetic reversals have already happened 171 times in the last 76 million years. Although not a certainty, scientific evidence supports the idea that reversal can happen abruptly rather than gradually. Evidence also exists that the reversal has already started. For example, earth's abrupt changing weather patterns and a rapid weakening of the magnetic field are precursors to a pole reversal. These phenomena are already happening.

26,000 Year Cycle

Our solar system makes a trip around our home galaxy, the Milky Way, every 26,000 years. The Milky Way spins around its center and at the center is believed to be a massive black hole. We are now at the end of one 26,000 cycle and the beginning of a new one. This transition to a new cycle allows an alignment between our earth and the galactic center which radiates a special energy to our planet. This energy could be a factor in the magnetic changes here on earth.

So What?

Although I enjoy learning about these scientific phenomena, what practical use is it? I'll tell you. Scientists have proven that our brains contain millions of magnetic particles. These particles connect, just as they do other animals, to magnetic fields on the earth, sun and the Milky Way. So, if Earth's magnetic fields are changing in the 2012 time frame, then we too are affected. One theory concludes that the Earth's magnetics play a key role in how we accept new ideas and changes in our life. With the reducing magnetic field, we become more open to change our ways. Since our magnetic field varies at different locations on Earth, we see evidence of this phenomenon. The lower the field the more change is acceptable, the higher the field the tendency is to hold to existing patterns and beliefs and little interest in change. The lowest level of the field occurs in the Mideast and in California. California has been the source of many new and improved ideas. The Middle East is certainly struggling for change. It appears that the way people handle the desire for change depends on their culture.

Changing your beliefs is the way to expand consciousness and this year is set up by the cosmos to enhance our ability to do just that. Recognize that our chance of humankind's transformation to a better life is highest during current times.

Now I step out beyond science to see what others are saying about the transformation of human consciousness and the year 2012.

Chaos theory tells us that in order to transform a system to a higher order, the old systems must fall. It is easy to see the breakdown of many current systems in our country and the world as a whole. The following systems represent some that are already falling apart:

- » Health care in the US is in trouble,
- » More and more corruption is being exposed in our government, showing how our leaders are abusing their power rather than serving the people,
- » In our global economy, the greedy and self-serving leaders are also being exposed to the public,

» The wars, terrorism, and the trouble in the Middle East, including the threat of nuclear warfare, are all jeopardizing our security and even our survival.

I also believe that we, especially those of us in the Western world, have been secretly manipulated by a relatively small number of people for centuries. These people have been and are now working to gain complete control of our planet with one global government, one global money currency and one global bank. Evidence of this intention is being uncovered now. Organizations such as The World Bank, The International Monetary Bank, The World Health Organization and others have been set up in preparation for this planetary control. These organizations are set up and controlled by a very small number of wealthy individuals. These families also control the major banks that control all large international corporations. It is known that international corporations have gained more power than the government of any country.

If we continue to abdicate our power to the few who are controlling our planet and our lives, we will lose our freedom and maybe even our survival. I believe we have the power to take back control of our lives.

I believe that our egos fool us into thinking we are much less than we are. This limits our consciousness and causes us to be unaware of the greater powers within us and hidden from us. Our limited consciousness leads us to believe that we are powerless to change the world. We have more power than our Western worldview allows. The world is suffering because we all think from a place of scarcity rather than abundance. I think the universe is infinitely abundant and we can tap into that unlimited abundance. I believe we have the resources to change the world. If enough creative people come together and accumulate a critical mass, a shift in human consciousness is probable. Or else we could devolve into destruction.

The good news is that destructive behavior is being exposed to the whole world. Powers that be are losing their foothold and their institutions are starting to crumble into the dust.

How do we evolve rather than devolve into oblivion? I believe we as individuals can do it by evolving ourselves during this time of transformation. The energy being transmitted from our galaxy's center, at this time, is affecting the earth and each one of us in an extraordinary manner. We are at the juncture between the 26,000 year cycles of the solar system's journey around the spinning Milky Way. As this new energy hits our bodies we experience changes in our perceptions, our brains and nervous systems and even in our DNA. Some of the physical changes that have been reported include:

- » Stiffness in the upper back, shoulders, and neck,
- » Transient joint and muscle pain,
- » Digestive issues,
- » Foggy headedness,
- » Short-term memory loss, and
- » Reality glitches.

Reality glitches are best described by this example: We enter another room to do something and find that it is already done, yet we have no memory of doing it. Conversely, we think we have done something and find that we did not do it, yet our memory tells us that we in fact did. I have certainly experienced reality glitches, loss of short term memory, and joint pain. I'm not saying that other causes are not at play, such as my age and my absent-minded professor-ness. Many others are experiencing and reporting the symptoms listed above.

Much of the above information comes from the book entitled *The Mystery of 2012, Predictions, Prophesies and Possibilities*[1]. For those of you interested in digging deeper into these ideas this book is a great resource.

Let's all work together not to save our planet, she doesn't need us to survive, but to save and evolve the Homo sapiens species, (that's us). The higher conscious humans evolving could have a new species name. Some proposed names include:

- » Homo universalis,
- » Homo spiritus,
- » Homo galactius,
- » Homo progressivus.

It is believed by many that the new species of humanity will be as different from us as we are from the Neanderthal.

The fact that some are talking about a new species indicates to me that the changes we face are enormously significant. However, I believe we need to change our thinking and our beliefs before the new species will emerge.

Chapter 7

Creating the New Science Worldview

Among the many changes from Newtonian Thinking to Quantum Thinking, I believe the two most important for us to focus on include:

- » Separation vs. unity, and
- » Exclusion vs. inclusion of consciousness

As covered in chapter 3, the Newtonian belief is that things are separate. This thinking focuses on *things* rather than relationships. Living entities are just things and the same as non-living matter. (One could argue that all matter is living). When we believe we are separate from other people and our home planet, we tend to focus on our differences rather than our similarities. When we believe in separateness our egos kick in to compare us to others resulting in "us vs. them" mentality. Fear takes control. This ultimately leads to hate, violence, war, greed, etc. Our treatment of the earth reflects the belief that we are separate from and not a part of our home. We treat the earth as a big rock with life on its surface, rather than a living organism with humans as part of the whole.

Our perception of separateness even at the macro level does not match reality. All living things interact with their environment and most of this interaction is invisible to the naked eye. As mentioned earlier, the 50 trillion or so cells in our bodies continuously die and new cells are reborn by interacting with its environment.

Nature evolves by focusing on cooperation more than competition. This concept challenges the Darwinian idea of "natural selection" which focuses only on competition. Our separateness thinking leads to competing with others in business and other settings. Our competitors become the enemy and we fight and try to destroy them. We often run our businesses as a military unit at war. Business jargon is often war related. For example, "win the battle but lose the war", or "let's fight our competition to win the business", or "let's hit the target".

The sense of scarcity from Newtonian Thinking also contributes to this competitive attitude. If we believe that our universe is abundant we wouldn't need to fight others for a piece of a fixed size pie. The pie increases through evolution and creativity so that all can thrive.

Consciousness

Newtonian thinking looks at *things*. Life is no different from inanimate objects. Quantum Thinking brings life into the equation through introduction of consciousness. The "observer effect," covered in Chapter 3, shows that an observer affects the status of matter. Consciousness even manifests matter (changes a wave to a particle). This puts us humans in a more powerful position to affect our world. Newtonian Thinking says our observations have no effect.

Bringing consciousness into science opens the door to help explain our relationship to our surroundings, including other people. With life included in physics, the study of biology changes from a materialistic viewpoint to a more holistic framework. That is, living things are more than molecules and include forces beyond Newtonian science.

Changing the Worldview

Why do we need to change our worldview? The current worldview based on Newtonian science is not working for us. Now is a turning point for Homo sapiens; we will destroy ourselves or evolve into a higher level of consciousness. In order to achieve the latter, we need to change our thinking.

The process we need, in my opinion, is:

Change beliefs > change thinking > change worldview > accelerate personal evolution of consciousness > join with others to obtain a critical mass > entire human race evolves to a higher level of consciousness > we survive and thrive without the major problems of today.

Change Beliefs

The first step in changing your beliefs is to believe you can change your beliefs. Without this, it is not likely that you can change.

The second step is to listen to your internal dialogue. Hear what you say to yourself. Often we discover we say negative things and usually these thoughts are repeated constantly throughout our lives.

The third step is to ask," Is this true?" If not true, you need to replace the negative thought with a true positive thought. First you notice the negative thought *after* you say it, then later you catch yourself *before* you say it. Then substitute a true and positive statement.

Change Thinking

Our thinking automatically changes as our beliefs change. Our actions change as our thinking changes.

Change Worldview

Our beliefs in total make up our worldview. As our worldview and thinking change from Newtonian thoughts to Quantum Thinking we can remove the limitations caused by the old thinking.

Consciousness

With the changing worldview we can open up to evolve to higher levels of consciousness. Meditation helps in expanding consciousness and this has been proven through many scientific studies. Scientists have documented changes in brain waves and other physical changes occuring at different levels of consciousness. It has been proven through many scientific studies that meditation helps expand consciousness.

Critical Mass

The next step is to join other likeminded people to generate a critical mass, or tipping point to change the consciousness of all humans. Deepak Chopra in an email provides the following quote:

> *"It takes as little as one percent of a population to create positive change, and I believe that if 100 million people underwent a personal transformation in the direction of peace, harmony, laughter, love, kindness, and joy, the world would be transformed".*

New World

As mentioned earlier, the changes in our world will be positive and extraordinary. Our species will evolve to bring forth a new species for humans. We live in a very unique and special time. We have the opportunity to change the world like never before. We can create a new species for humans and a new planet for all life.

Let's all work together to change our world into a peaceful and nurturing place to live.

Bibliography

Introduction and Chapter 1

1. Braden, Gregg. Deep Truth: igniting the memory of our origin, history, destiny, and fate: Carlsbad, CA. Hay House, Inc. www.hayhouse.com. 2011.
2. Marciniak, Barbara, Path of Empowerment: Pleiadian wisdom for a world in chaos. Makawao, HI. Inner Ocean, 2004.
3. Ibid.
4. Lipton, Bruce H. PhD. The Biology of Beliefs: unleashing the power of consciousness, matter & miracles. Carlsdad, CA: Hay House, Inc.: www.hayhouse.com. 2008.
5. Dyer, Wayne, W Excuses Begone: how to change lifelong, self-defeating thinking habits. Carlsbad, CA. Hay House, Inc. www.hayhouse.com. 2009.
6. Marciniak, Barbara, Path of Empowerment: Pleiadian wisdom for a world in chaos. Makawao, HI. Inner Ocean, 2004.
7. Braden, Gregg. *The Spontaneous Healing of Belief*: shattering the paradigm of false limits. Hay House, Inc: www.hayhouse.com,2008.

8. Ruiz, Don Miguel and Don Jose Ruiz with Janet Mills. The Fifth Agreement: a practical guide to self-mastery. "A Tolec Wisdom Book", 2010.
9. Marciniak, Barbara. Path of Empowerment: Pleiadian wisdom for a world in chaos. Makawao, HI. Inner Ocean, 2004.

CHAPTER 4

1. Braden, Gregg. *Deep Truth*: igniting the memory of our origin, history, destiny, and fate: Carlsbad, CA. Hay House, Inc. www.hayhouse.com. 2011.
2. Friedman, Norman. *Bridging Science and Spirit*: common element in David Bohm's Physics, the perennial philosophy and Seth. St. Louis, MO. Living Lake Books,1990.
3. Schrodinger, Erwin. *My View of the World*. Cambridge University Press,1964.

CHAPTER 5

1. Chopra, Deepak. The Book of Secrets: unlocking the hidden dimensions of your life. New York. Harmony Books <www.crownpublishing.com>, 2004.
2. Russell, Peter .The Global Brain: speculation on the evolutionary leap to planetary consciousness. Los Angles. J.P. Tarcher, Inc. 1983.
3. Briggs, John and F. David Peat. Seven Life Lessons of Chaos: timeless wisdom from the science of change. Harper Collins Publishers, Inc. 1999.

CHAPTER 6

1. Miscellaneous authors, The Mystery of 2012: predictions, prophecies, & possibilities. Boulder, CO., Sounds True, Inc. 2007.

CPSIA information can be obtained at www.ICGtesting.com
Printed in the USA
LVOW041113301112

309490LV00005B/13/P